biting at the PANORAMA, LEICESTER SQUARE.

37. Aiguille du Plan.
38. Glacier de la Blaitière.
39. Glacier du Plan.
40. Chalets des Blaitières.
41. Glacier des Pélerins.
42. Aiguille du Midi.
43. Mont Maudit.
44. Les grands Rochers Rouges.
45. Sommet du Mont Blanc.
46. Le grand Plateau.
47. Rochers des grands Mulets.
48. Glacier des Buissons.
49. Montagne de la Côté.
50. Glacier de Tacconay.
51. Montagne de Tacconay.
52. Glacier de Gria.
53. Montagne de Gria.
54. Dome de Goûte.
55. Aiguille de Goûte.
56. Aiguille du Bionnassay.
57. Aiguille du Miage.
58. Pierre Ronde.
59. Le Bonhomme.
60. Mont Lacha.
61. Le Breven.
62. Passage de la Cheminée.
63. Chalets de Plampra.
64. Sentier de la Croix de Flégère.

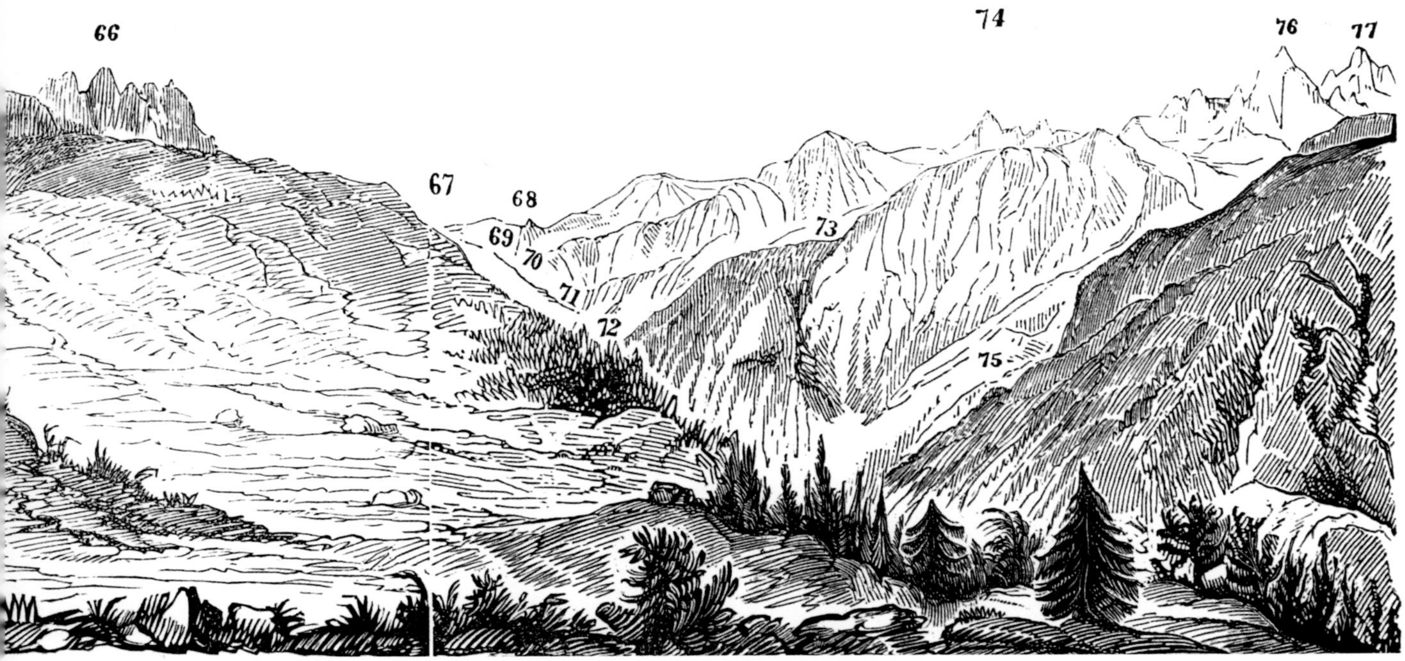

72. Hameau du Tour.
73. Glacier du Tour.
74. Aiguille du Tour.
75. Glacier d'Argentière.
76. Aiguille du Chardonnet.
77. Aiguille d'Argentière.

Flora of Chamonix

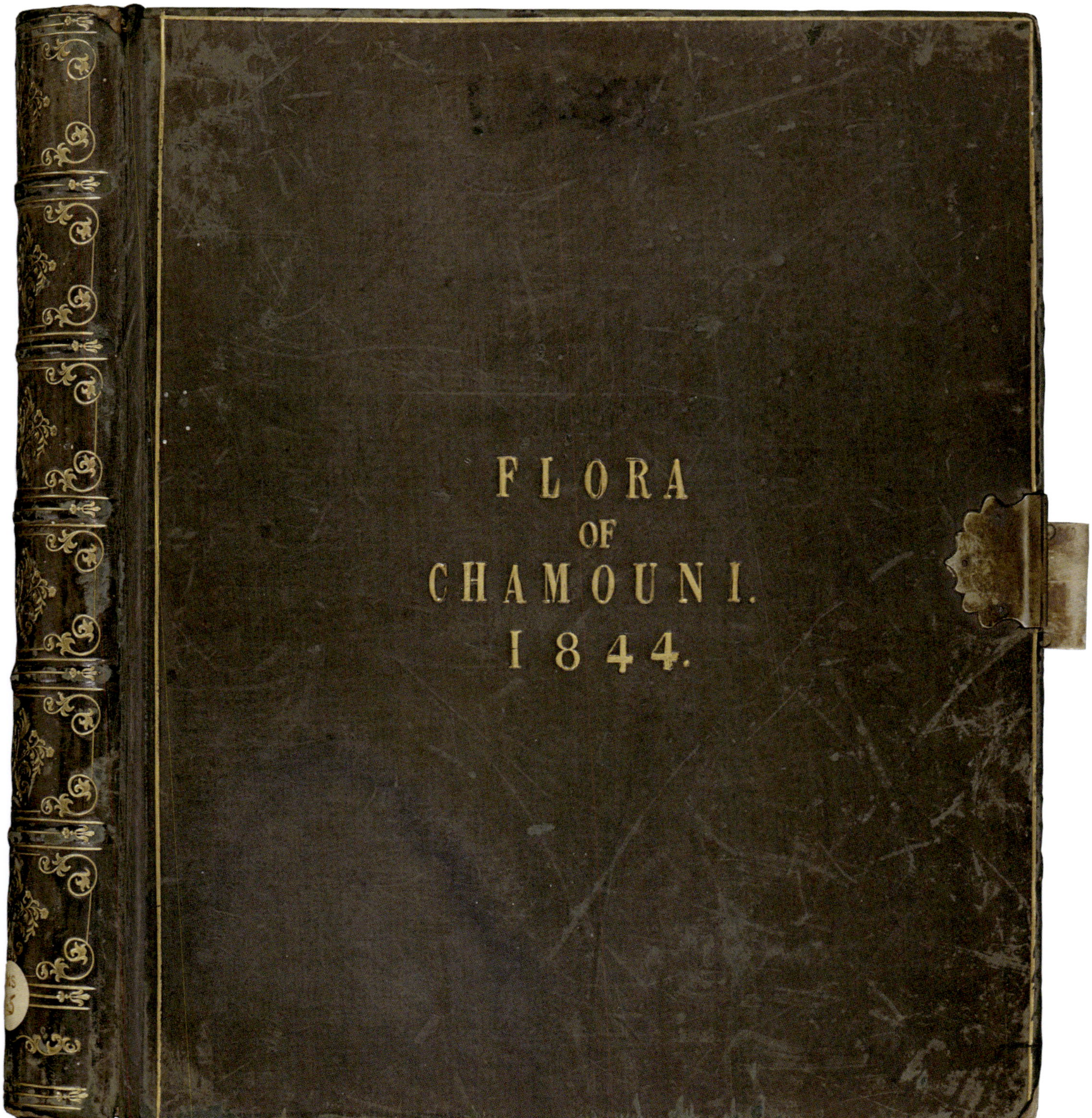
FLORA
OF
CHAMOUNI.
1844.

Flora of Chamonix

JOHN RUSKIN

EDITED AND INTRODUCED BY
DAVID S. INGRAM
AND STEPHEN WILDMAN

VOLUME ONE: FACSIMILE

PALLAS ATHENE

Dedicated to the memory of the Ruskin scholar

JIM DEARDEN

who knew so much and helped so many

MAYANTHEMUM BIFOLIUM

Muguet a deux feuilles.

The two specimens. a & b. were gathered on the 5th June. among the woods which clothe the lower slopes of the Mont (du Reposoir?) which rises on the right in approaching Cluse from Bonneville by the old road. The plant seems to love sloping banks. growing chiefly beside the hollow paths cut or sawn out of the gravelly soil by the dragging down of the pinetrunks. These. to the number of 11 or 12. are bolted together by the extremities and are drawn down the steep sandy slopes by a single horse.

Vide large collection. Vol 3. 238.

DRABA AZOIDES

Drave faux. aizoon

This plant was growing, June 5th. in enormous quantities. about the Lac Benit and on the bare turfy summits under the great precipice of the Mont du Reposoir associated chiefly withe the Soldanella. Potentilla, & large bell gentian. Vide notes in journal. I found it afterwards frequently in the same situations - but never in such profusion

SOLDANELLA ALPINA.

This, certainly the boldest, though not the most ambitious, of Alpine flowers, was growing, June 5th in profusion about the Lac Benit, & more sparingly half way down. Above the Flegere, June 8th I found them finer, though less numerous – one of these is preserved at p. 9. Again, on the crest of the ridge which separates the valley leading to the Col de Berard, from the Val d'Entragues, and which overhangs the latter in an awkward precipice, facing north, in crossing some slopes of half frozen snow – I saw the purple head of this flower thrust up through the snow, which it had pierced, though an inch thick.

a. b. Soldanella Alpina. c. d. Gnaphalium dioicum.

GLOBULARIA. CORDIFOLIA.

It was growing in masses which made the ground purple. all about the Lac Benit. associated with the Soldanella. their delicate purples opposed to the gold of the Potentilla. and the common crowfoot. But its colour is now entirely gone. It was I think, much fresher in appearance when I first took it out of its envelopes. and has lost what remained to it of colour by exposure to the light. What vestige of hue remains to it, is however its original one. it has not turned purple from red or blue. as is the case with many plants

GENTIANA ALPINA.

I seldom gathered the gentians - finding it impossible to preserve their colour. but I found them this day growing in such extraordinary masses on the very top of the Montagne Benit. that I gathered one or two for memorials. I have given away all but this; and I might as well have parted with this too - for it has not only lost its colour. but shrunk to one half its original size. and is now a very slander on itself and its companions. I never afterwards in all my Alpine rambles, met with this gentian either in such perfection or profusion - and owing to the early season. the flowers having just burst, their colour was intense beyond imagination. As the season advances, the colour diminishes greatly in purity & vigour: Fine specimens are only to be found by climbing high. to spots where their blossoming has been late. Thus a zone of bright blue. followed by one of pale blue. ascends the mountain sides as the summer advances.

The two specimens were gathered by my Mother, from the meadows under the Nant d' Arpenaz. while I was absent on the Mont du Reposoir.

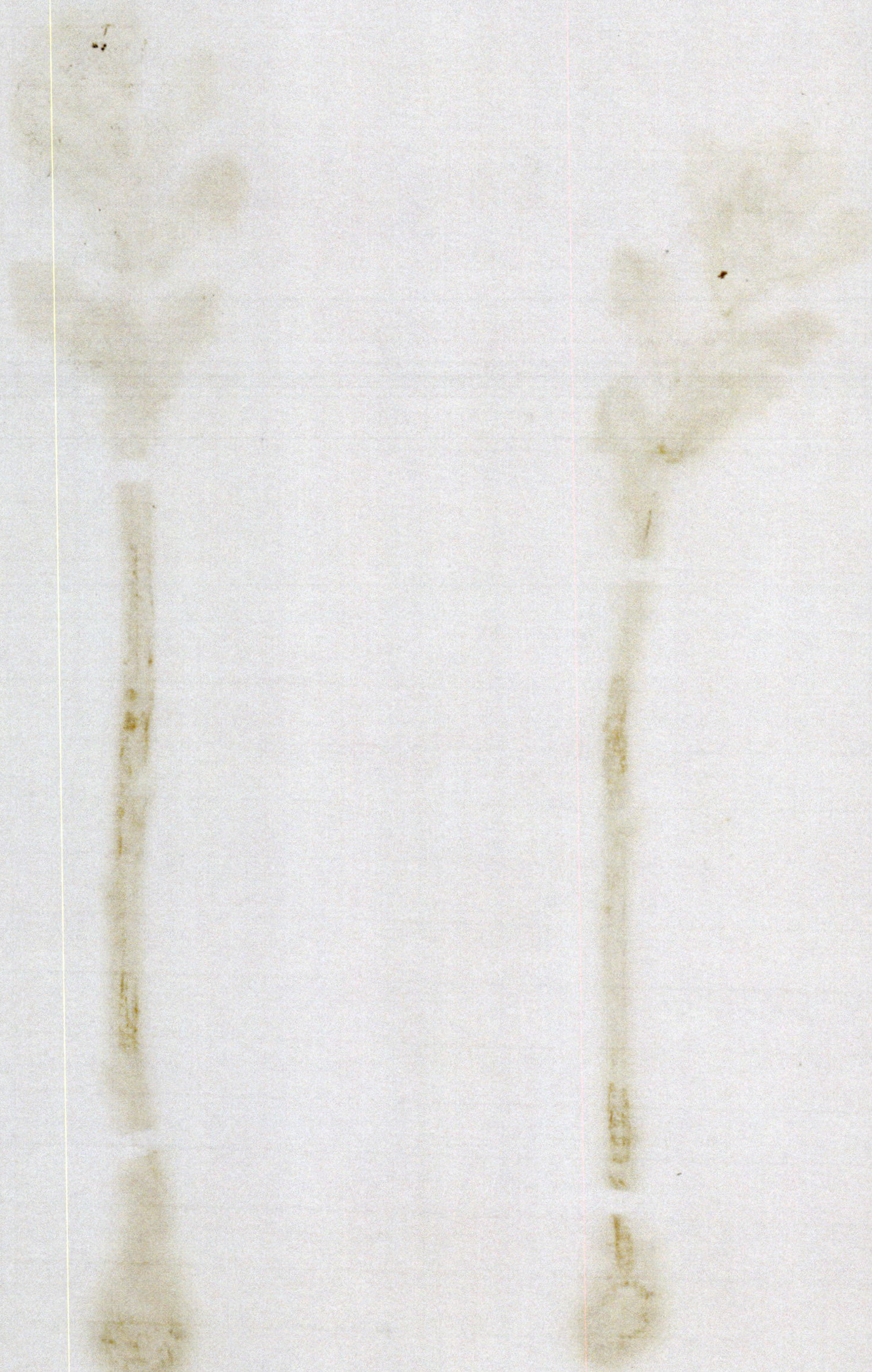

ORNITHOGALUM FISTULOSUM.

All the specimens were gathered on the 8th of June. while ascending the steep pasturages above the pines, which lie between the Flegere and the northernmost of the Aiguilles Rouges. It is an abundant flower in all such situations.

GEUM MONTANUM

Benoite de Montagne.

Gathered the same day. and abundant everywhere. but I think in a zone somewhat above the region of the preceding plant.

Gathered the same day, merely as a reminiscence – (for the plant is everywhere abundant on the higher hills) of the place where we rested – lying plunged over head and ears in masses of the Alpine rose, under a grey bit of rock, which kept off the sun, and from a cleft in which tinkled a rivulet which we stopped with moss into a little crystal pool, and drank like horses; and then slept to the singing of it; the vision of the Mont Blanc, a great white throne, floating before ones eyes as they closed, entangled among the burning purples of the rhododendron.

RANUNCULUS PYRENAEUS

Very frequent in the Alpine pastures at four or five thousand feet above the sea. It should have come before the preceding plant. having been gathered a little lower with the Geum Montanum. p. 7. The Soldanella I found at the same time growing in considerable abundance. and larger than I ever saw it elsewhere.

These are exceedingly bad specimens of the flower. having been gathered above its proper zone; from the very highest rock I reached that day; near the foot of the aiguilles rouges; and at about the same elevation as the top of the Breven; I could not get farther; for the rocks beyond were inaccessible except inch by inch; and our time was nearly exhausted. The specimen of the orchis nigra below (vide p.) were gathered coming down. from the wet meadow - the last bit of sloping ground in descending from the Flegere.

POTENTILLA AUREA

I imagine these two specimens, a. and b. to be both the same flower. though the blossom in one is so much more developed. at the expence of the strength of the plant. This smaller one, b. was gathered on the Mont. du Reposoir. June 5th. and was there associated with the leaves c. both plants growing in great quantities high and low. I cannot find the name of the plant c. and I have lost the ~~to~~ name of the locality from which I obtained the other potentilla. a.

ALCHEMILLA VULGARIS. .C.

a.

b

c.

I found this plant frequently in damp places among the deep wood mosses; but these specimens were gathered between the Prieure and the source of the Arveron in our quiet walk on the morning of Sunday. June 9th just at the place where the bridge over the brook leads to the shattered block of granite & unfinished obelisk. The gnaphalia below. were from the steep bank near the Cascade des Pelerins. & were gathered the same evening – in my solitary walk. I took them for a rarer flower and had a scramble for them. swinging myself up by the pines; and there was a noble glimpse of stormy light down the valley at the time. which I should have lost. had it not been for the chase of the flower; which gave me command of it from a higher point.

SAXIFRAGA CUNEIFOLIA

This saxifrage, common everywhere, was growing in great luxuriance (June 9th) in the lower woods of the Pelerins, whence I gathered these in my evening walk. But all the form of the flower is now lost.

SAXIFRAGA. ROTUNDIFOLIA.

These plants I found June 10th just under the gallery of the Tete Noire, growing among stones by a meadow streamlet. I never met with the flower in such perfection again.

PRIMULA FARINOSA

Below the saxifrage places, we came to some steep meadows, sloping to the brink of the eau noire - where I found this primula. and I imagine the grasses below - Coutet was always gathering grasses; as if I had wanted a hayrick instead of a Flora. and these are probably his selection.

These grasses were from the same meadows: but are common everywhere. They were intended to form part of a series of grasses: but I was not again among the low meadows and got only hill plants.

In returning, the same afternoon, June 10th I walked as far as the turn up the valley to the col de Berard. There I waited for Coutet, whom I had lost sight of, among the rocks between the Chalets de Berard and the Tete noire road. The meadows underneath were always covered with the bell gentian and globe ranunculus: the latter peculiarly fine. but I only gathered these grasses, for fear of crushing my primulas and saxifrages.

When Josef came up the same afternoon he brought me these gunce as if they were rare. I kept them therefore, though they seemed to me one of the most frequent meadow plants.

These mosses were gathered June 11th in walking to the Chapeau through the Arveron woods. together with a plant of the Pinguicula. vid P. 30. which was growing luxuriantly in the green bed of a little running stream. The moss is very frequent in the woods.

This moss was gathered in the same spot as the preceding. but it occurs in far greater abundance among the dry stones left by the great avalanche beneath the cascade des Pelerins. growing in deep & thick masses.

SAXIFRAGA COTYLEDON. L

Saxifraga aizöon Jay.

Gathered 11th June. in my scramble with Coutet up the moraine of the Glacier des Bois. just under the Chapeau. about the place where the granite joins the Buet slate. This saxifrage was growing on the patches of soil which had lodged in the hollows of the moraine.

SAXIFRAGA ASPERA

Saxifrage rude.

Just beneath the chapeau, and at the edge of the moraine of the Glacier des Bois there are some projecting rocks. (granite) en place, which are nearly bare of vegetation and consequently exhibit, better than any that I know, the effect of the glacial action. On the tops of these, however, this, & the other hardy saxifrages grow in greater perfection than I ever found elsewhere.

SAXIFRAGA MUSCOIDES

Saxifrage mousse.

From the same rocks as the preceding species.

POLYGALA ALPINA

Polygale des Alpes.

This lovely plant grows luxuriantly on the same rocks as the preceding saxifrages. I do not think it is uncommon. but have not found it elsewhere. It was not on the barren parts of the rocks. but just where the rich wood vegetation began to appear.

Gathered in the marshy ground under the cascade des Pelerins as we returned from our stormy morning walk. June 12th

Both specimens were gathered on June 12th from the short-grassed steep pastures under the Cascade des Pelerins. It is a common flower, but I never saw it of so vivid an orange as at this spot. Its colour was quite lost in drying.

This plant (a large shrub) grows luxuriantly in the clefts & shady hollows under the two small cascades east of the Casc. des Pelerines. I did not meet with it elsewhere.

ORCHIS NIGRA

Also from the meadows near the cascade des Pelerins June 12th. It is not unfrequent. & its vivid point of black is a striking feature of an Alpine pasture.

ANDROSACE CARNEA. L.

I only found this lovely plant once, during all my Alpine rambles, and that was on the 14th of June, in ascending the steep, short-turfed slopes above the Montanvert in order to reach the base of the aiguille Charmoz. It was growing not far above the Montanvert, being when fresh, of a most pure, pale, life like rose colour; and growing in bright delicate clusters of starry flowers -five leaved. But it lost its colour in drying, totally.

TUSSILAGO ALPINA.

This plant I found the same day. June 14th and I am nearly sure, on the high & barren ridge of rock which runs from the foot of the Petit Charmoz down beside the Glacier de Blaitière; and on which that day I rested, before my hot struggle up the steep snow of the glacier. It might have been from the lower ridge - but I think not. at all events it was growing in a barren spot. islanded among snows. at 4000 ft above the Prieuré.

THALICTRUM AQUILEGIFOLIUM

It is exceedingly common in all the rich pastures on the lower slopes of the hills I preserved it only on account of the elegance of its leaf. It was gathered on the 14th of June. in the steep meadows just opposite the Prieure - on the left of the cascade. as I came down from the glaciers of the aiguille Charmoz.

PINGUICULA ALPINA.

This plant is the most difficult to preserve of any Alpine flowers I have tried. It is full of a fatty liquid. (hence its name) which causes it to take weeks before it dies - even under the heaviest pressure - and commonly it decays before it can be effectually dried. There is a marked difference, however, between those of the high & low localities. the three uppermost. 1. 2. 3. were from the col above the Val d'Entragues, gathered June 17th (compare p. 3): there associated with the Soldanella. and growing in the mossy hollows which were wet with the meltings of the snow. at about 2500 ft above the Prieuré. These are evidently stronger in stem & firmer in constitution, (keeping their white colour) than the one below. which was gathered June 11th. in the wood; going to the Chapeau. in a very damp spot. (vide p 19.) and which lost its colour. and all appearance of form. long before I could dry it.

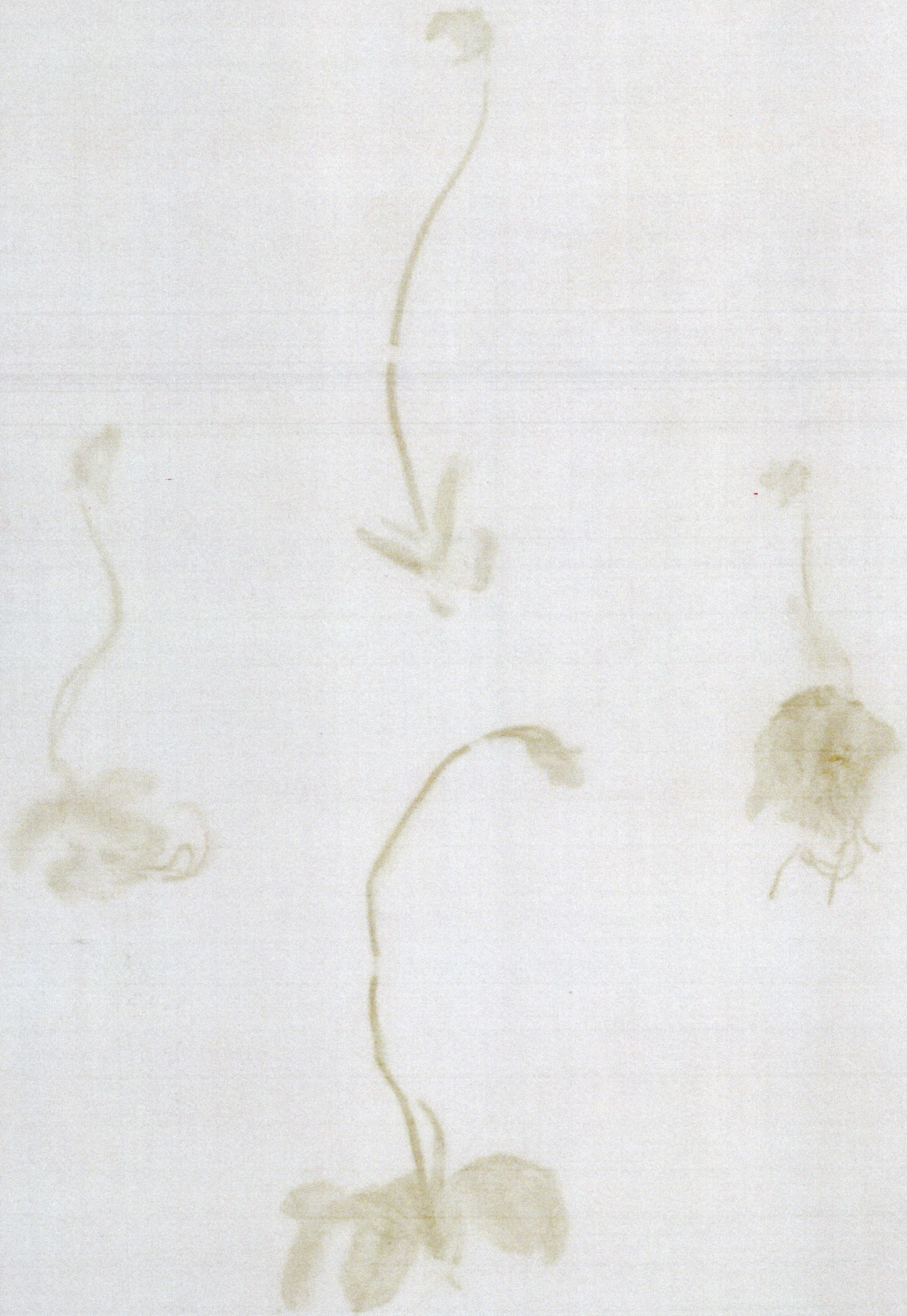

This plant I found growing here and there sparingly in the woods on the side of the Breoen. in my evening walk on the 20th June. It grows very low. one or two blossoms on a stalk. Those below are from the same locality. The wood slopes steeply. to the south east. and the soil is comparatively parched & bare of flowers.

VERONICA URTICIFOLIA.

The largest and darkest had lost its number when I unpacked it; but is I believe, (the other is certainly) from the same woods as the preceding. A different group of flowers appears to me to grow on this part of the Breven, from those of any other locality in the valley.

GNAPHALIUM DIOICUM.

These specimens, which were the finest I ever found, are from the open heights which I reached the same evening, after passing through the wood; and about half way up the Breven.

Half way up the Montagne de la Côte; June 21. we sat down to rest on the first bit of ground we had come to, that afforded more than a step for the foot and a ledge for the hand. It was at the top of a small cascade which slid for some yards trickling and glittering, over a rounded slope of smooth rock. and then fell from the edge of it in divided streamlets, nearly down to the Glacier de Taconay. In the middle of the stream. and in the crannies of the rock, grew this plant. associated with the flowers of the R. glacialis (vide p 45) preserved with it. and which were the first I had ever seen.

SAXIFRAGA. BRYOIDES

This delicate little flower I have marked only as from the side of La Côte. June 21. nor do I recollect its exact locality.

Also from the side of La Cote. June 21. I was so fatigued when I came down that evening that I fell asleep in writing: and never noted the exact localities.

[illegible] glabratum 108

On the 22nd of June in climbing to the foot of the aiguille du Midi, I found this plant, either in the thicket at the bottom, near the glacier des Bossons which I rather think, or else in the open meadows just above.

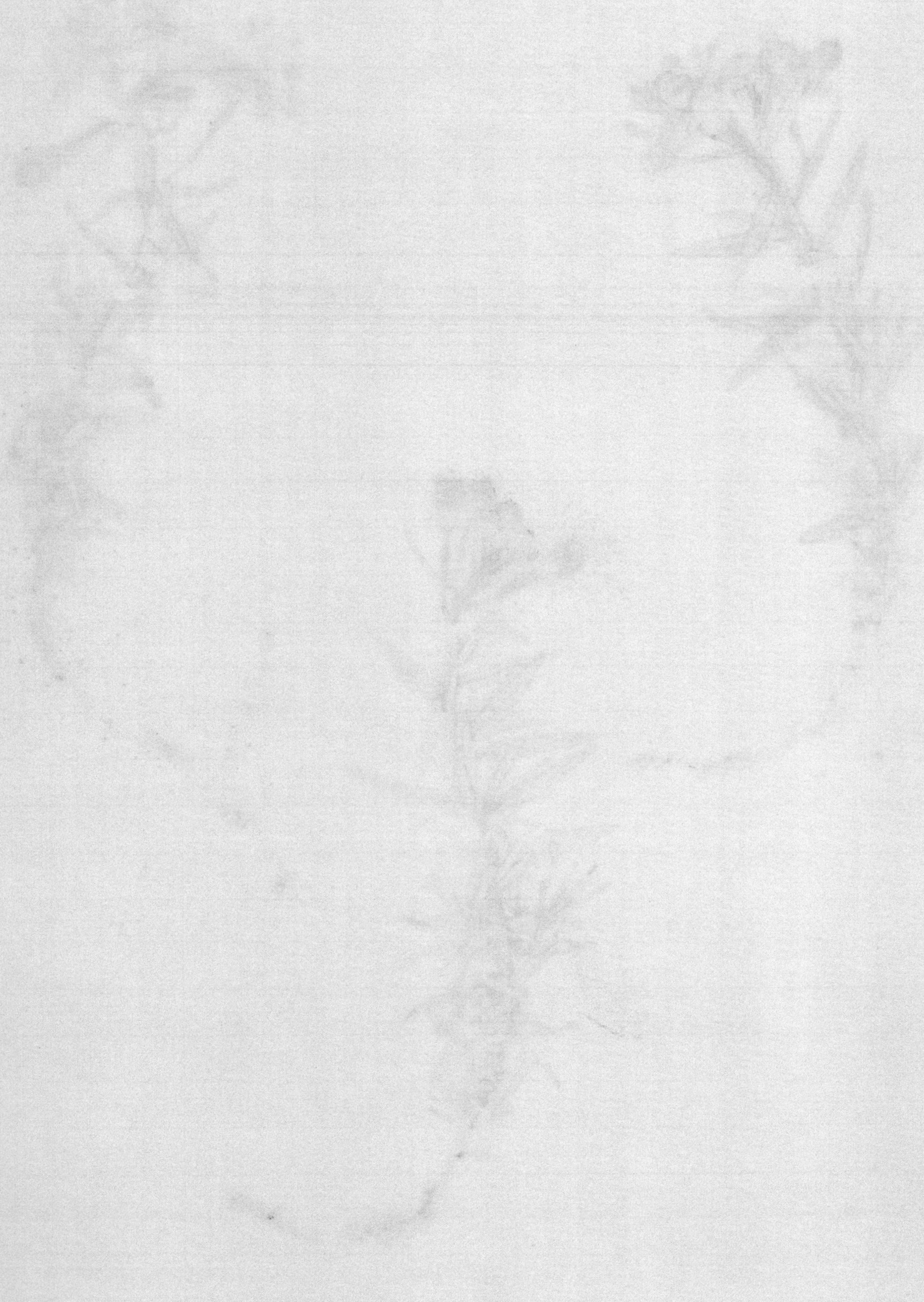

LINARIA. ALPINA.

this lovely plant. though I do not imagine it is uncommon, I have yet only found twice among the Alps. the first time, this day. as we ascended from the border of the glacier des Bossons, through the couloir, to the shoulder of the aiguille. The watercourse was the steepest bit of continuous ascent I climbed in Chamonix For twenty minutes we could not loose hold with our hands without danger. and there was some chance also of stones coming from above. so that we hardly stopped to breathe. the soil too was dry & dusty, and loose for the foot. & the only redeeming point of the whole was, that this flower. then new to me. -was growing in large bunches of the most intense purple - all about the crannies of the loose soil. The second time that I found it was on the very highest turn of the Simplon road. the evening that I walked up from the village to see the sunset on the Aletsch horn. ( ) there also growing on a dry - dusty soil to which its roots hardly adhered. There I found two species, one like this, with the vivid orange centre. the other having this orange replaced by a pale lilac. In both these situations the plant was growing at more than 6000 ft above the sea.

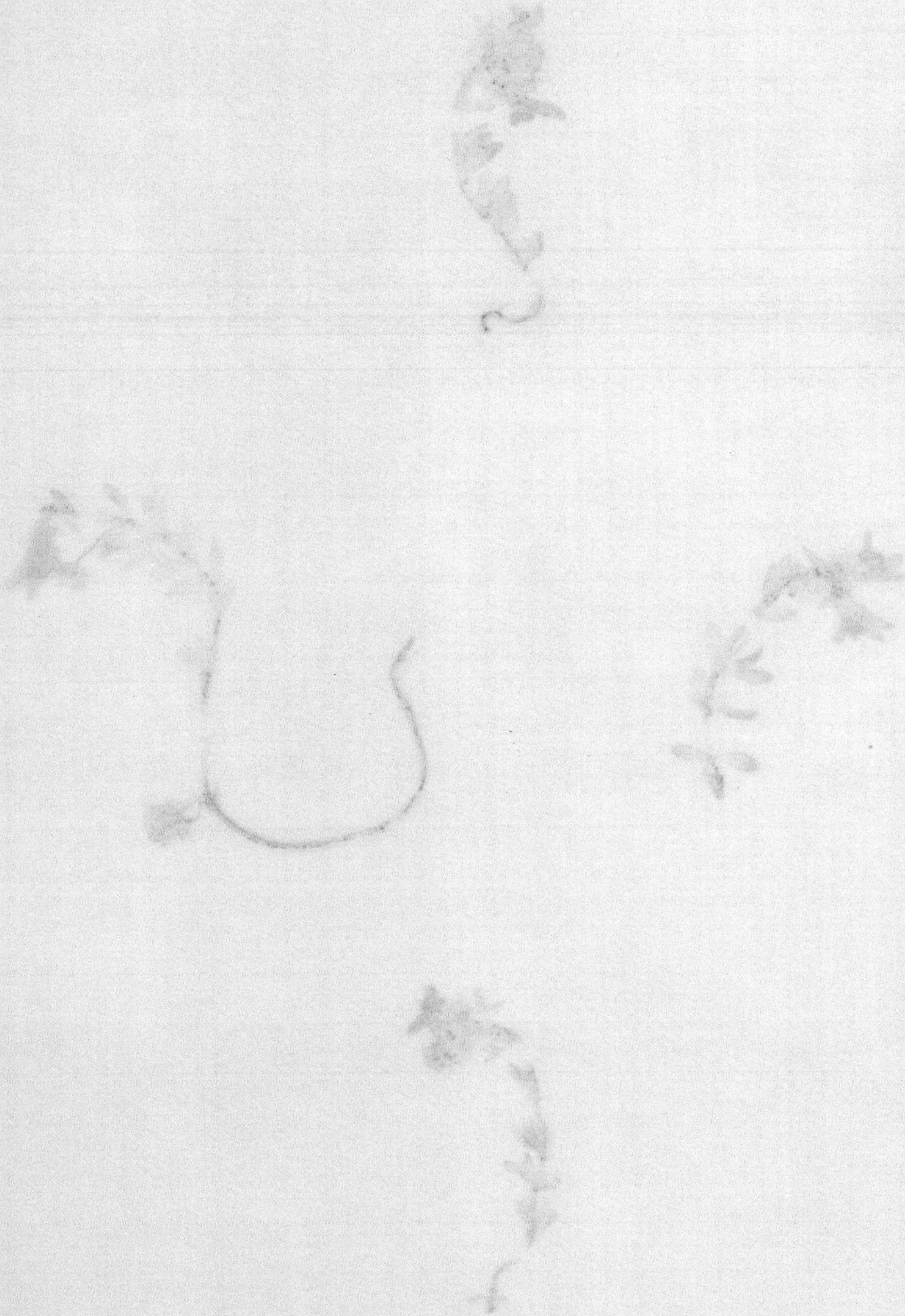

This I believe to be an exceedingly rare plant. I never saw it but once. and then I could not get at it myself. but Coutet scrambled down into the chasm & brought it up for me. indeed he was the first to point it out. It was in a rocky cleft. just at the base of the aiguille du Midi; near the point of entering on the glacier in ascending Mont Blanc: about 500 ft above the couloir in which I gathered the preceding plant. Its colour when fresh, is an intense smalt - a mixture of lake and ultramarine; exquisitely beautiful.

This plant is certainly one of the hardiest of the Alps. I found it growing in large bunches, June 22nd. among the loose blocks of the uppermost moraine of the glacier des Pelerins. at a height of nearly 9000 ft above the sea, close to the ice. and associated only with the Ranunculus Glacialis. Comp. p. 45.

ASTRANTIA MINOR.

This plant is a very characteristic feature of the lower rock foregrounds of Chamouni. this specimen, a sufficiently bad one, was gathered June 24th by the road side in the ravine by which the Arve leaves the valley. Its leaf occurs in a foreground study made on the Montanvert. The Veronica, from the same place, is the most frequent species of Chamouni. It occurs in greatest perfection about the Tete Noire.

This plant was gathered June 24th in the flat marsh which forms the foreground of my sketch with the pines, near les Ouches. I preserved it because Coutet affirmed it to be the best remedy known, against fevers.

These specimens were gathered June 27th close beside the bridge which forms the lowest part of the pass of the Tete Noire; and from which the road ascends to the gallery.

THYMUS ALPINUS

From the same spot as the preceding species.

Gathered on the rocks, by the side of the path, a little nearer Chamonix than the preceding plants. The rocks were the "poudingues de Valorsine" of Saussure. and the situation close to the arch which marks the boundary of Savoy & Switzerland.

RANUNCULUS GLACIALIS

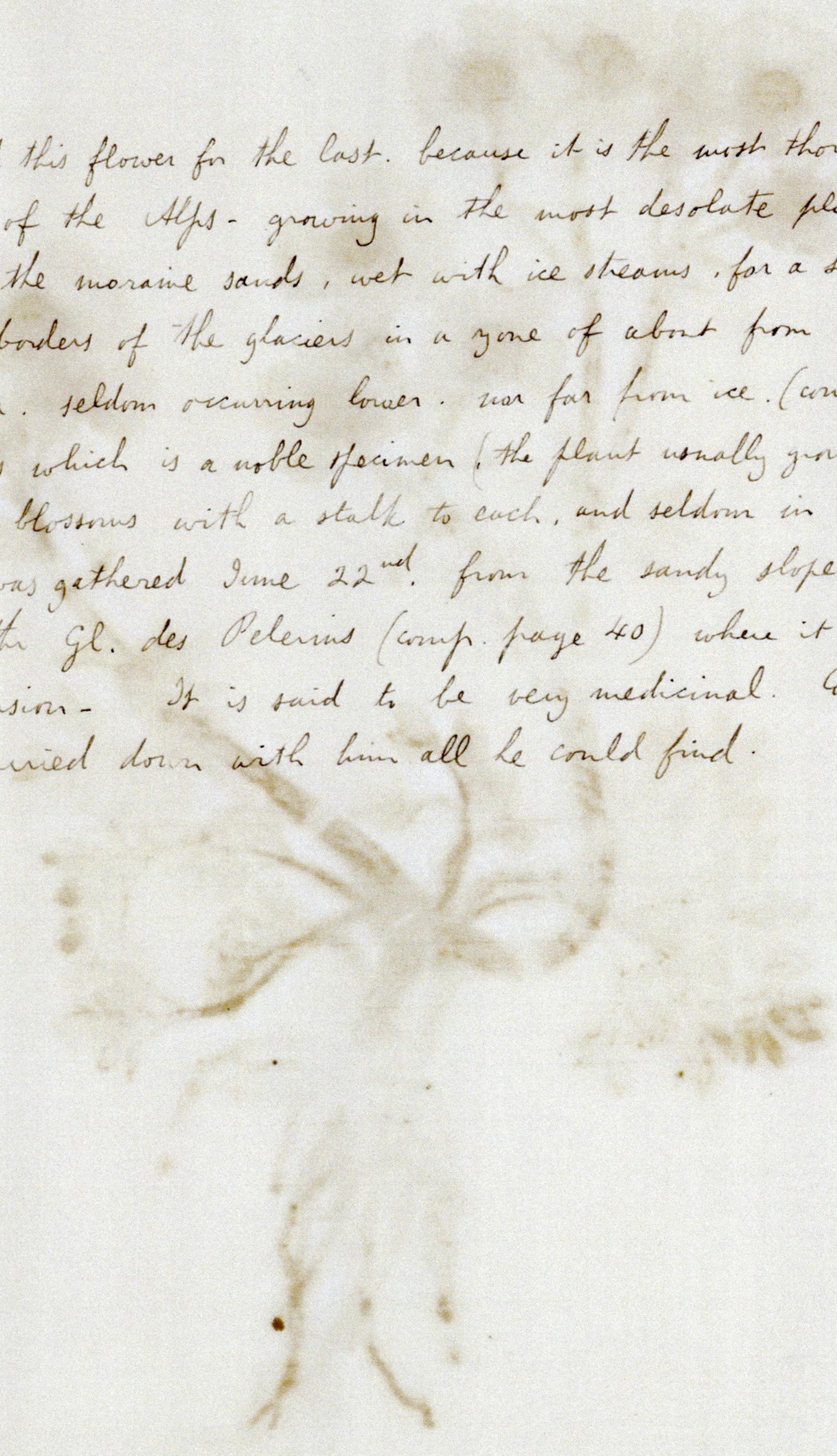

I have reserved this flower for the last. because it is the most thoroughly characteristic of the Alps - growing in the most desolate places. with nothing but the moraine sands, wet with ice streams, for a soil. It haunts the borders of the glaciers in a zone of about from 5 to 9000 ft above the sea. seldom occurring lower. nor far from ice. (compare however p. 34.) This which is a noble specimen (the plant usually growing low & in bunches of blossoms with a stalk to each, and seldom in this branch like form) was gathered June 22nd. from the sandy slopes of the moraine of the Gl. des Pelerins (comp. page 40) where it was growing in great profusion - It is said to be very medicinal. Coutet always gathered & carried down with him all he could find.

FLORA OF CHAMONIX by John Ruskin,
edited by David S. Ingram and Stephen Wildman

Volume I: Facsimile ISBN 978 1 84368 231 8
Volume II: Commentary ISBN 978 1 84368 232 5
Set: ISBN 978 1 84368 233 2

First published 2023 by Pallas Athene (Publishers) Ltd
2 Birch Close, Hargrave Park, London N19 5XD

Book design and layout by Alexander Fyjis-Walker

Printed in England

1. Aiguille Verte.
2. Aiguille du Dru.
3. Glacier du Nant Blanc.
4. Le Chapeau.
5. Mer de Glace.
6. Glacier des Bois.
7. Aiguille de Bochard.
8. Rochers des Mottets.
9. Aiguille du Moine.
10. Petite Jorasse.
11. Glacier de Léchaud.
12. Grand Jorasse.
13. Les Periades.
14. Le Tacul.
15. Le Jardin.
16. Montanvert.
17. Sentier de la Filia.
18. Source de L'Arveiron.
19. Hameau des Bois.
20. Sentier du Montanvert.
21. Hameau des Mouilles.
22. Route de Martigny.
23. Hameau des Prés.
24. L'Arve.
25. La Prieuré de Chamounix.
26. Chalets sur le Rocher.
27. Valley of Chamounix.
28. Hameau des Pélerins.
29. Cascade des Pélerins.
30. Pont Piralotta.
31. Hameau des Buissons.
32. Mont Mallet.
33. Le Géant.
34. Aiguille des Charmoz.
35. Glacier de Crepon.
36. Aiguille de la Blaitière.

65. Cabane de la Croix de Flégère.
66. Aiguilles Rouges.
67. Chaine Septentrionale du Valais.
68. La Genousi.
69. Rochers de Croix de Fer.
70. L' Hospice du Col de Balme.
71. Chalets de Charamillan.